IMAGES
of America

MARINSHIP

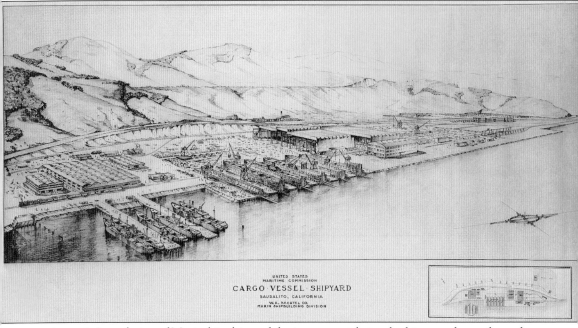

An artist's rendering of Marinship shipyard depicts an aerial view looking northwest from above Richardson Bay. (Author's collection.)

ON THE COVER: This view looks north toward the bow of the *William A. Richardson* on launch day, September 26, 1942. The first ship launching at Marinship was only six months after shipyard construction had begun on a vacant tidewater mudflat. One of the shipyard's largest launching crowds watched the ship slide into the waters of Richardson's Bay only 91 days after the keel was laid on an unfinished shipway in an unfinished shipyard. (Courtesy of the Sausalito Historical Society.)

IMAGES of America
MARINSHIP

Eric J. Torney

Copyright © 2018 by Eric J. Torney
ISBN 978-1-4671-2977-0

Published by Arcadia Publishing
Charleston, South Carolina

Printed in the United States of America

Library of Congress Control Number: 2018935198

For all general information, please contact Arcadia Publishing:
Telephone 843-853-2070
Fax 843-853-0044
E-mail sales@arcadiapublishing.com
For customer service and orders:
Toll-Free 1-888-313-2665

Visit us on the Internet at www.arcadiapublishing.com

This book is dedicated to the hundreds of thousands of nameless and unselfish men and women, minorities and otherwise, young and old, who patriotically responded to the call to duty to build ships and planes and help win the war against German and Japanese aggression during World War II.

Contents

Acknowledgments		6
Introduction		7
1.	Introduction to Marinship	11
2.	Marinship Is Built	23
3.	Marinship Is Operated	47
4.	Marinship Characteristics	61

ACKNOWLEDGMENTS

This book, about an incredible shipyard built during World War II in an almost unbelievably brief time, is inspired by an accidental discovery of a trove of historical material. On first viewing the disorganized and amazing collection of visual, video, and written materials, I became smitten. I swore to return and to organize it into some sort of an account of what was then a legend.

The shipyard had always been a fascination for me. I remember being driven by it when I was very young and asking my father what it was. He would say that it was a shipyard—and not much more. My father had been a B-24 bomber pilot during World War II, so he was in Europe for most of the time the shipyard operated and had little appreciation for it.

As I sifted through this archive, I became astounded by the accomplishments of the shipyard. This book tells the story as I have slowly become aware of it. It is truly the "World's Most Amazing Shipyard" in many different ways—not just for what it accomplished by building ships, but also by how it affected the local community of Sausalito, the overall San Francisco Bay Area, and the outcome of the war, as well as how it affected the labor workforce and modern society in general.

By creating this book, it is hoped that an awareness of the importance of this shipyard can be communicated to those who appreciate such things and that this appreciation will help to preserve and commemorate the remains of the shipyard, much of which still exists. Marinship is the most complete remaining emergency World War II shipyard. At least 18 of the original 29 major buildings or structures remain, many of which have been recycled to serve as office buildings or studios.

The Sausalito Historical Society is thanked for allowing access to and use of its historical materials and resources. Its oral interviews with Marinship workers can be found at https://archive.org/details/sausalitohistoricalsociety.

Thank you to Frederick Penn, National Park Service ranger at the Presidio of San Francisco, for his help identifying the entertainers pictured on page 119.

Images in this book appear courtesy of the Sausalito Historical Society unless identified as otherwise in the caption.

INTRODUCTION

THE MARINSHIP SHIPYARD – WORLD'S MOST AMAZING SHIPYARD – 1942–1945

The Marinship shipyard was located in Sausalito, California, just north of the Golden Gate Bridge. It was operated as an emergency shipyard during World War II to build urgently needed ships to defend the free world from assault by Germany and Japan.

The shipyard was built on a vacant mudflat, a small hill, and an abandoned railroad maintenance yard. The mudflat was filled by bulldozing the small hill for fill, and the railroad maintenance yard was modified for the new use. Altogether, there were about 210 acres spread along a one-mile-long site needed for the shipyard. Thousands of men and women worked at the shipyard.

As the beginning days of American involvement in World War II dawned, it soon became apparent that there would be a need for many new cargo ships and oil tankers. An emergency shipyard program was proposed by the US Maritime Commission. A total of six entirely new shipyards were requested during March 1942. Marinship was one of those emergency shipyards.

The US Maritime Commission needed contractors to build and operate those shipyards. Recent government building projects done as part of the Works Progress/Work Projects Administration effort to employ American workers during the Depression in the 1930s used a group of contractors who called themselves Six Companies. These six companies (actually eight of them) banded together to assist each other, each offering a specialty service best provided by the individual organization. As they banded together to accomplish those unbelievably large WPA projects, one of them proposed the name Six Companies, to associate themselves with the six Benevolent Tong Associations of San Francisco, California's Chinatown. Hoover Dam, Grand Coulee Dam, Bonneville Dam, the San Francisco to Oakland Bay Bridge, and the Golden Gate Bridge were some of these projects. These projects had such grand and huge scope and scale that no individual construction company then in existence could have ever accomplished it alone. The W.A. Bechtel Company, based in San Francisco, was one of these companies. It had endured the Great Depression and was then establishing itself as one of the major construction firms on the West Coast. It had assembled an organization of many talented and capable individuals who knew how to work together and how to accomplish difficult and challenging construction projects. W.A. Bechtel died in 1933 while working on a construction project in Siberia, but by 1942 his sons Steve and Kenneth were very capably managing the company.

Ken Bechtel was soon to become the president of Marinship Corporation. He was the guiding force and spirit behind the success of the Marinship shipyard.

The W.A. Bechtel Company had begun in the shipbuilding business in 1939, in a partnership with Six Companies contractors. Among the Six Companies, besides the W.A. Bechtel Company, was the genius industrialist Henry J. Kaiser. The British government had placed an order for 60 cargo ships, and the Kaiser No. 1 Yard in Richmond was built to supply these badly needed maritime vessels.

Subsequent to this initial involvement in shipbuilding with Henry J. Kaiser, the W.A. Bechtel Company developed its own shipyard in Southern California to build ships. This shipyard, Calship, was built starting February 1941 on Terminal Island in Los Angeles Harbor. It became the second-largest emergency shipyard operating during World War II. Of the total 461 ships Calship contributed to the World War II defense effort, 306 were Liberty ships and the rest were other types of cargo ships and tankers.

Soon after the United States became involved in World War II, immediately after the Japanese attacked Pearl Harbor, it became apparent that the German U-boat Wolf Packs and Japanese submarines were becoming very efficient at sinking cargo ships. In the first five months of 1942, a total of 268 ships were sunk: 17 in January, 36 in February, 60 in March, 68 in April, and 87 in May. It was very clear that ships were being sunk much faster than they could be built. An American emergency shipyard building program was promptly begun.

On March 2, 1942, the W.A. Bechtel Company received an urgent telegram from the US Maritime Commission's Admiral Land pleading for the W.A. Bechtel Company to build a new shipyard somewhere on the West Coast. Key W.A. Bechtel Company administrators, including company president Steve Bechtel, Ken Bechtel, and a few others, immediately had a telephone conference. It was decided that Ken Bechtel would be the most effective leader for this new shipyard. On Ken's review of potential Bay Area shipyard sites, that afternoon it was quickly determined that most of the possible sites for a shipyard were already occupied by shipyards and that Marin County should be investigated because it had not been so industrially developed as other Bay Area locations.

On the night of March 2, 1942, Ken Bechtel called on some friends living in Marin County. They agreed to tour potential sites the next day. Ken and a few of his key men visited and evaluated local Marin County sites, with the Sausalito site selected that day. The Sausalito site was selected because it had unused railroad capacity and a deepwater channel adjacent to vacant and available land. The Northwestern Pacific Electric Commuter Railroad and Ferry system had been closed down because of the opening of the Golden Gate Bridge, leaving the rail line and deepwater channel unused and available.

During the afternoon of March 3, Ken Bechtel called Admiral Vickery, vice chairman of the US Maritime Commission, and described his recommendation. Admiral Vickery responded favorably and asked Ken to prepare a proposal and to come to Washington to present it. The next few days were occupied by producing the requested proposal. Ken flew to Washington, DC, and on March 9 presented his proposal to Maritime Commission officials. Ten minutes into the presentation, Ken was told to go ahead and build the shipyard. That afternoon, in a telephone conversation with Steve Bechtel, it was decided that Steve would begin preparing plans and get started with preliminary work on the site. Work to build the shipyard was begun only one week after the initial request for it had been received and before a formal contract had been awarded.

Ground was broken on March 28, 1942. Work progressed as rapidly as the extraordinarily qualified and competent W.A. Bechtel Company could manage. Labor and construction machinery was summoned from within the resources of the company. As the germ of a shipyard began to become visible, the Calship Corporation was tapped for a manager of operations. William Waste was transferred to Marinship and began organizing to build ships. Key experienced shipyard workers were selected to come to Marinship from Calship to get things started.

Not only labor came from Calship. Prefabricated ship parts that could be loaded on railcars were made and delivered to the developing Marinship site. On June 27, only three months after ground had been broken, as soon as the faint form of an incomplete shipway became apparent on what only months before had been vacant land and tidewater, a keel for the first Marinship ship was laid on it. Prefabricated ship parts transported from Calship were immediately added to the keel. Three months after laying the keel, on September 26, 1942, a mere six months after ground had been broken, the first ship, the *William A. Richardson*, was launched. *Amazing!* A ship launched from an incomplete shipyard that only six months before had been a vacant mudflat.

Ken Bechtel had the talent, the management ability, the people-motivating skills, and the support of a proven construction organization to do all of this. He did not stop there.

As the shipyard formed and as the first ship was being built, the US Maritime Commission observed that Marinship was the most efficient and most productive of all other emergency shipyards. As the construction contract was being produced in March, in its great wisdom, the Maritime Commission had asked for slight design changes that would allow larger ships than the originally requested Liberty ships to be built. A few days before the *William A. Richardson* was to be launched, Admiral Land came to Marinship to announce that he was going to change its production to tankers. The change in production was made because Marinship had demonstrated extraordinary and effective shipbuilding capabilities. From then onward, Marinship built tankers instead of Liberty ships.

Marinship built 15 Liberty ships and 78 tankers. From the date the *William A. Richardson* was launched, a new ship went down the ways and into the bay at Marinship on average every 11 days. Among the chronology of the ships built at Marinship is the still-standing record for the world's shortest construction time for a tanker: the *Huntington Hills*, built from laying keel to delivery, ready for service, in only 33 days. Comparable tanker records include the *Swan Island* (60 days), *Alabama* (79 days), and *Sun Ship* (90 days). Under Ken Bechtel's direction, new production methods and new labor management techniques vaulted Marinship from one of the newest shipyards built to become one of the finest, most advanced, most modern, most competent, and most productive shipyards in the entire United States. Marinship could hold its own against any other shipyard, even those established long ago.

The tankers built at Marinship were among the most sought after during and after the war. They were the fastest, had the largest and most powerful propulsion system, were among the most well constructed, and were considered the most beautiful. After the war ended, Marinship tankers were most favored for conversion to peacetime uses. Many were used as tankers, but some were modified to become the first-ever container ships. The surplus Marinship tankers were so well built and well regarded that several tankers could be cut into sections and up to three original tankers reassembled to be one supersized ship.

Besides the ships built at Marinship, there were other features of the organization that enhanced and immortalized its fame. The labor and management arrangement was among the best of any manufacturing facility operated during the war. Finding reliable labor was challenging. The draft took many skilled men needed as soldiers and sailors. This slowed production. An alternative labor source was needed. Minorities, rarely before allowed in union jobs, and women were recruited. These recruits were not skilled as shipyard workers; in fact, many had never even dreamed they could build ships. But Marinship developed effective training programs and implemented mass production assembly strategies, which enabled it to achieve amazing production records. There were major labor union problems as a result, but at Marinship these were handled with superior administrative skill. Marinship became known as one of the most effectively integrated workforces anywhere. The workforce problem resolutions developed at Marinship can be identified as being one of the sources of the integrated workforce we have today: men and women and minorities working side by side, together, in relative harmony, to accomplish the same goal.

It was Ken Bechtel's leadership and management style that inspired all of this to happen. Active in local civic organizations, he was a family man, commissioner of the Boy Scouts in Marin County, and a philanthropist. Without him, it is doubtful whether Marinship would have been able to distinguish itself as admirably as it did.

This is an aerial view looking north at the Marinship site. The photograph, which was donated to the Sausalito Historical Society by Steve Bechtel, shows the shipyard in full production. Bechtel proudly displayed the image in his office for many years. On the lower left is a cluster of buildings surrounding the Outfitting Docks, which are easily identified by the eight tankers at the docks. At the Outfitting Docks, ships are loaded with the supplies needed for operation at sea. Above the Outfitting Docks are the areas of the shipyard where steel plates are fabricated into ship parts to be assembled into a ship on the shipways. Bridgeway, then known as State Highway, runs along the west side of the shipyard. Marin City, built to house shipyard workers, is beyond at the top of the picture.

One

Introduction to Marinship

The remnants of the shipyard today have been at rest for many years. On March 6, 1946, the US Army Corps of Engineers took over the shuttered shipyard site. The 20,000 workers had been laid off in stages, beginning with victory in Europe. The need for ships had been fulfilled. Ships already under construction and significantly complete were finished. Those ships not yet substantially completed on the shipways were abandoned and scrapped. On the occasion of victory in Japan, the shipyard was put to sleep. Only a few hundred workers were retained to inventory the materials and equipment and to guard the site. The US Army Corps of Engineers auctioned off the parts of the site that it did not need for its operations. The story of what this amazing shipyard was those many years ago can now be told and appreciated.

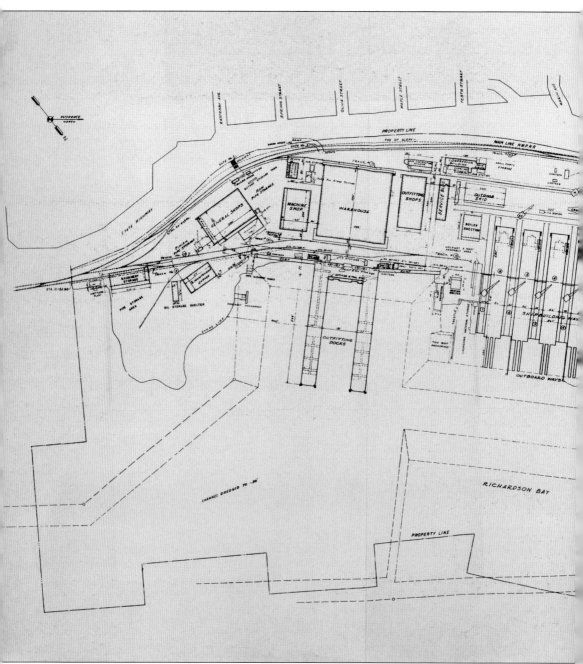

This Marinship plot plan shows the entire shipyard site. All major buildings and work areas are identified. The shipways are centered in the shipyard. Craneways (tracks on which cranes moved) are shown. More than one crane was necessary to move objects around on the site and then onto a shipway, relay fashion. At the top of the map is the street today known as Bridgeway. It is

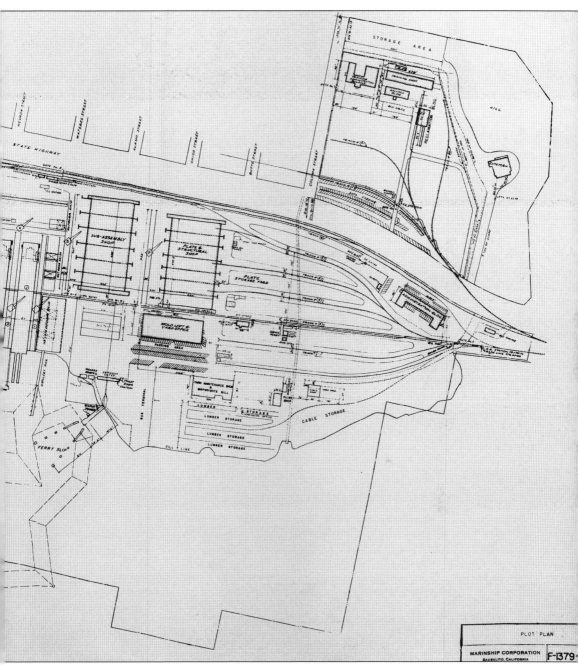

located along the west edge of the main part of the shipyard. A main rail line is located parallel to and near Bridgeway. Internal rail lines are shown within the perimeter of the shipyard site. Richardson Bay is at the bottom edge of the map. In the bay are dotted lines, which represent limits of dredging needed to provide adequate depth of water to float the ships.

This panoramic image shows the shipyard near the end of operations. The view is to the north from near the south end of the shipyard west of today's Bridgeway. Beyond and above the house in the foreground is the Warehouse, which is just below the shipways. To the right of the Warehouse are the Outfitting Docks. The Subassembly Building, near top left in picture, is concealing the Plate Shop and the Yard Office and Mold Loft. Note the cars for workers parked along Bridgeway. Gates for entry to the shipyard are located near the ends of the parked cars. One gate is in the foreground, near the end of the guardrail extending to the right from the closest parked car. A small guardhouse is barely discernible against a tree at the top of a driveway.

It all began on December 7, 1941, when the Japanese surprise attack on Pearl Harbor induced a reluctant United States of America to enter the hostilities of World War II. Incensed, the Americans entered the war with a vigor and determination never anticipated by the Japanese invaders. (National Archives.)

As the beginning days of involvement in World War II dawned and the reactions from Americans progressed from initial shock to reality, it was observed that Japanese submarines and German U-boat Wolf Packs were becoming very efficient at sinking cargo ships. Something had to be done to counter this threat. The first step was to organize convoys to try to protect the ships. But this was not working too well. More ships and better protection for the ships were necessary. Building more ships was the more effective, more easily achieved approach, but they needed to be built quickly and the United States did not have a viable and vibrant shipbuilding industry. New, efficient shipyards were needed.

2 MARCH 1942
W. A. BECHTEL CO.

IT IS NECESSARY IN THE INTERESTS OF THE NATIONAL EMERGENCY THAT THE MAXIMUM NUMBER OF EMERGENCY CARGO VESSELS BE COMPLETED PRIOR TO DECEMBER 31 1942.

THE COMMISSION BELIEVES THAT IT HAS UNDER CONSTRUCTION AND PLANNED CONSTRUCTION SUFFICIENT TO ACCOMPLISH MAXIMUM TONNAGE REQUIRED FOR THE CALENDAR YEAR 1943. THE PARAMOUNT PROBLEM IS COMPLETED TONNAGE IN 1942. THEREFORE, COMMISSION IS REQUESTING THE MEMBERS INDIVIDUALLY OF THE SIX COMPANIES, INC. GROUP TO EACH SUBMIT A SEPARATE PROPOSAL FOR CONSIDERATION OF THE COMMISSION AS FOLLOWS:

1. PROPOSED SHIPYARD SITE, ESTIMATED COST OF THE CONSTRUCTION OF THREE OR FOUR WAY FACILITIES FOR THE CONSTRUCTION OF EC-2 CARGO VESSELS, SUCH SITE AND PLANT LAYOUT TO BE LOCATED IN ONE OF THE WEST COAST PORTS IN WHICH YOUR ORGANIZATION COULD OPERATE TO THE BEST ADVANTAGE.

2. THE PROPOSAL SHOULD INCLUDE THAT YOU WOULD BE WILLING IN THE INTERESTS OF NATIONAL DEFENSE TO PROCEED WITH THE CONSTRUCTION OF SHIPS UNDER THE TERMS AND CONDITIONS OF THE CONTRACTS COMMISSION IS PRESENTLY OFFERING FOR THE CONSTRUCTION OF EC-2 VESSELS.

3. YOUR PROPOSAL SHOULD INCLUDE A SCHEDULE OF DELIVERIES BEARING IN MIND THAT THE MAXIMUM NUMBER OF COMPLETED SHIPS IS DESIRED PRIOR TO DECEMBER 31, 1942.

4. IN VIEW OF THE FACT THAT THE CONSTRUCTION OF FACILITIES AND SHIPS HAS ALREADY BEEN CONSUMMATED BETWEEN THE COMMISSION AND YOUR GROUP AS A WHOLE WE ARE NOW RELYING ON YOU INDIVIDUALLY IN THE INTERESTS OF THIS EMERGENCY TO CONTRIBUTE YOUR INDIVIDUAL ORGANIZATIONS IN THESE INDIVIDUAL YARDS FOR THE PURPOSE OF SECURING COMPLETED SHIPS IN THE PRESENT CALENDAR YEAR. THE EMERGENCY DEMANDS ALL WITHIN YOUR POWER TO GIVE YOUR COUNTRY SHIPS.

PLEASE ADVISE BY WIRE DATE WE MAY EXPECT YOUR SUBMISSION AT THIS OFFICE.
E. S. LAND, U. S. MARITIME COMMISSION

No copy of the original telegram from the US Maritime Commission that was sent to the W.A. Bechtel Company is known to exist. But the text of that message has been documented. A facsimile of that urgent message is shown, with most of the message under a grey overlay to highlight the most important sentence of the message. The W.A. Bechtel Company responded immediately and sent Ken Bechtel on an assignment to find a site and to build and operate a new six-shipway shipyard. Sausalito was chosen as the site for this shipyard. The Sausalito site selected had available land, an abandoned railroad line already in place, and access to a deepwater channel.

Ken Bechtel is pictured on a launching platform in the Marinship shipyard. (Courtesy of the Bechtel Corporation.)

The peaceful shipyard site is pictured on March 23, 1942, five days before construction was begun to convert it into a shipyard. The viewpoint is looking south from Waldo Point. Today's Bridgeway is in the foreground. The railroad tracks will be relocated to be closer to Bridgeway, and the mudflat will be filled to create land several hundred feet beyond the railroad causeway.

This is the shipyard site, seen in a view looking south toward Pine Point days before construction was begun.

This view of the shipyard site faces north from Pine Point, days before construction was begun.

Here is another south-facing view of the shipyard site from Pine Point, days before construction was started. The abandoned railroad maintenance yard is shown, with much of the original equipment and facilities cleared. Most of the railroad trackage has already been removed.

This view of the shipyard site faces north toward Pine Point from the south end of the railroad maintenance yard, days before construction was to begin.

This is the shipyard site in a view looking northeast from a hill above Sausalito. Visible are Pine Point and the partially cleared railroad maintenance yard before construction had begun. Strawberry Point and Richardson Bay are in the background. The intersection of today's Bridgeway and Spring Street is near the center right section of the image.

Pine Point is pictured in this view looking in a northerly direction from near the intersection of Filbert Avenue and Marie Street. This image is comparable to the one on page 11 but is from before construction began. The railroad maintenance yard is shown before its removal. All of Pine Point to the right of today's Bridgeway would be removed down to shipyard site level and used for fill to create the shipyard.

Pine Point is demolished. Families living in houses on Pine Point were given about two weeks notice to move. The families generally moved with patriotic enthusiasm in support of the war home front effort. Many houses were moved to nearby sites. A number of the houses were moved to Spring Street or to other nearby vacant lots. The houses that could not be moved were demolished.

Two

Marinship Is Built

Construction began before a contract was signed because of the urgency of the need for cargo ships. The W.A. Bechtel Company allocated internal resources to get work started immediately. Machinery and men were put on the job as soon as the US Maritime Commission gave the go-ahead to build the shipyard. From the Calship shipyard, William Waste was brought in to be general manager. As the top man, he was responsible for having plans drawn, building the shipyard and the ships, and making sure everything ran smoothly and efficiently. Ted Panton was assigned to be construction manager. Ted was on the site directing the construction activities. But more workers and machinery would be needed. Other contractors who had worked with the W.A. Bechtel Company were engaged. Work progressed slowly at first, but as more construction workers and machinery became available work became hectic. After specialty construction jobs were completed, those workers no longer needed for shipyard construction became shipyard workers. Only months were needed to accomplish what would normally have been expected to take years. This was war. The workers were angry, and they were determined. They accomplished things in unbelievably swift times, overcoming immense obstacles to achieve their goals.

One of many Earth Moving Days was May 14, 1942. Pine Point is almost gone, demolished for fill to create the shipyard site. The Administration Building construction has begun (left center). Pile drivers are hard at work placing 25,952 pilings driven for supporting major structures to be built on filled land. Today's Bridgeway runs across the photograph from lower center left to upper center right. Today's Gate 5 Road starts at lower center left.

EARTH-MOVING. This picture was taken on May 14, 1942. Pine Point Hill was almost gone, the driving of 25,952 timber piles had been started a week earlier, filling operations were well advanced, and the Administration Building (left) was being erected.

FROM THE SAUSALITO HILLS Marinship presented this appearance on May 28, 1942, two months after ground breaking. Pine Hill has been leveled and pile drivers are working on the coffer dam and foundations for the ways, while the Warehouse takes shape in the right foreground.

Marinship is pictured from Sausalito hills on May 29, 1942, two months after construction began. The last remnant of Pine Point is the first row of trees to the left of the rectangle of sheet pilings in the bay. The sheet pilings are placed to allow pump removal of tidewater where the six shipways are to be built. The Administration Building is rising up, now two floors, at far left center.

The south part of a cofferdam is being put into place to allow shipway construction on land that was originally mudflats or tidewater. Pilings to be driven are seen floating in the water. Pilings driven for Shipway No. 6 are seen at center right, and to their left are pilings for Shipway No. 5. Remnants of Pine Point are visible in the foreground as it is being demolished.

This is a view looking northeast from Sausalito hills. The top end of Shipway No. 6 has been built, and Shipway No. 5 is beginning to become recognizable. The Warehouse Building is nearly fully roofed, and the Outfitting Shops Building, immediately to the left of the Warehouse, is fully formed and enclosed. The rectangular form of sheet pilings, placed to facilitate shipway construction, is clearly evident dead center in the picture.

This grand panorama shows the shipyard as it is being built. This most historic image shows history being made. On the far left, the Administration Building is almost complete, if it has not already been occupied. Thousands of pilings are laid out, ready to be driven. The six shipways will be built at center rear. Shipway No. 6, on the right and quite unfinished, is almost ready to receive the keel for the *William A. Richardson*. Stacked in racks in front of Shipway No. 6 are

This most historic panorama shows a closer view of the shipways under construction. Shipway No. 6 is seen to the right, being readied to receive the first sheets of keel steel to be laid for the *William A. Richardson*. Planks that will be part of Shipway No. 6 are pictured on the right, laid on support beams. Shipway No. 5 is being formed on the left. Shipway No. 5 support beams are being positioned on pilings. Pile drivers are busy at work building the foundation for the lower part

ship components that were fabricated at the Calship shipyard, on Terminal Island in Southern California. The ship parts were loaded onto railcars and delivered to the unfinished Marinship site, stacked in order of assembly, ready to efficiently build the first ship to be launched. On the far right is the Warehouse Building, under construction.

of Shipway No. 5 and for Shipway No. 4. The cofferdam is in place, allowing work to proceed at areas where tidewater previously predominated. Mobile cranes are doing the heavy lifting until the tracked craneways for huge whirley cranes can be built. The closest mobile crane occupies what will soon be a whirley crane craneway.

The Marinship site is pictured under construction in mid-June 1942, just prior to keel laying for the *William A. Richardson*. Shipway No. 6, where the Richardson's keel will be laid, occupies the foreground in this picture. Five more shipways will be built in the space beyond the foreground. Waldo Point, in front of Mount Tamalpais, is seen beyond and to the left of the closest pile driver.

This is the view looking south from near the middle of the shipyard site. Shipway No. 6 is seen in the center background, just above the roof of the pickup truck. The Warehouse (with the light-colored roof) is pictured under construction just beyond Shipway No. 6. Railroad tracks shown at center bottom are on the original Northwestern Pacific Railroad right-of-way. The tracks will be moved a few feet to the left.

The foundation for the Compressor Building is pictured at center foreground. The foundation for the Service Building is at center right. Outdoor skids are at center left. Shipway No. 6 is seen forming just beyond the mobile crane at center. Railway cars loaded with prefabricated ship sections are being unloaded onto outdoor skids at center left. Construction is progressing at a frantic and urgent pace.

Plant Protection Building Foundations

This view shows the plant protection building foundation, located just to the south of Shipway No. 6. The top end of Shipway No. 6 is nearly complete on the right near upper center. What may seem to be a disorganized construction site, with concrete formwork lying helter-skelter in the foreground, is actually a refined and well-managed construction project operating under extreme efficiency. Within days, the formwork will be in place and concrete trucks will arrive.

The Yard Maintenance Shop and Shipwright's Mill foundation is shown under construction. Shipways No. 5 and No. 6 are seen at top center background. The forest of scaffolding supports surrounds the ship being built on Shipway No. 5. Shipway No. 6 is presumed to be beyond, with the three huge whirley cranes beside the ships under construction between them.

The Administration Building, shown several days after its dedication and on the day after the keel for the *William A. Richardson* was laid. The Administration Building has been occupied, and construction activities are being directed from within it. Today's Bridgeway runs from lower left to center right and then up over the remaining part of Pine Point toward downtown Sausalito.

Sub Assembly Building during construction circa June 1942 looking Northeast

The Subassembly Building's steel frame is shown under construction. This view is looking to the northeast from a location to the east of today's Bridgeway, near Nevada Street. In this building, sheets of steel that will be cut and formed in the Plate Shop, pictured being built beyond and to the left of the Subassembly Building, will be assembled into recognizable ship subassemblies. Ship assembly work on the crowded and congested shipways is minimized.

Administration Building and shipyard site during construction frm Waldo Point circa June 1942

Shipyard site during construction is seen from Waldo Point in mid- to late July 1942. The steel skeleton of the Plate Shop is at center. The Administration Building is at center left. The Yard Office and Mold Loft is above the Administration Building. The Subassembly Building's steel skeleton is beyond the Plate Shop's skeleton. To the left of the Subassembly steel skeleton, the hull of the *William A. Richardson* is being formed.

Yard Office and Mold Loft

The Yard Office and Mold Loft is shown under construction. This photograph's subject is nearly impossible to identify by its content. The only visible clue is the small sign just to the left and in front of the mobile crane working inside the building.

Yard Office and Mold Loft

The Yard Office and Mold Loft is shown under construction in this northwest-facing view. This photograph's location was only identifiable by comparing it to a number of other images to confirm what it shows. Counting the number of bays being built and the number of floors was the confirming information. Two floors are shown, but there are posts extending another level higher.

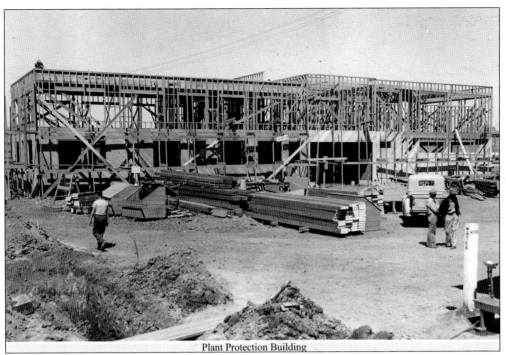
Plant Protection Building

The Plant Protection Building is shown under construction. A small sign near the center of the photograph gives this obscure building under construction its identity.

Machine Shop under construction

The Machine Shop is pictured under construction. The historic identification of this building misidentifies it. A detailed study of its components and comparison to other images reveals its true identity.

This late 1942 view is looking north from the site where the General Shops Building will be constructed. The Machine Shop is left of center. The idyllic shipyard location features Mount Tamalpais. One Marinship tanker, the flagship of the Pacific Fleet, was named after the mountain. Tankers are named only after waterways, so an unnamed creek on the mountain was found and named Tamalpais Creek. The *Tamalpais* was in Sagami Bay commemorating the Japanese surrender.

Mold Loft and Yard Office under construction. The Mold Loft and Yard Office, at 107,630 square feet floor area, was one of the largest buildings at Marinship. As Yard Office it was one of three headquarters for Marinship management. The other two were the Administration Building at 3030 Bridgeway (for the top executives including Ken Bechtel and Bill Waste) and the Construction Offices, located right at the Ways. The dramatic steel framing to the right of the Mold Loft is the beginning of construction of the Plate Shop.

The Mold Loft and Yard Office is shown under construction. This view is to the southeast. At 107,630 square feet, the Mold Loft and Yard Office was one of the largest buildings in the shipyard. As Yard Office, it was one of three headquarters for shipyard management. The dramatic steel framing to the right of the Mold Loft and Yard Office is the beginning of construction of the Plate Shop.

A COFFER DAM IS COMPLETED to permit construction of the ways on pilings being driven into the mud. A large part of the year rests on an "inverted forest" of more than 30,000 of these pilings.

The cofferdam allows shipway construction in an area normally inundated by tidewater. The forest of pilings shown is only a fraction of the 25,952 driven to support shipyard buildings built over bay mud originally as deep as 90 feet. The tops of some of these pilings can be seen today when bay tidewater is low. Many humps in today's parking lots reveal other pilings long ago covered over and forgotten.

Pictured here is a keel-laying ceremony on June 27, 1942, three months after construction of the shipyard had begun. From left to right are Steve Bechtel, president of W.A. Bechtel Company and vice president of Marinship; Ray L. Hamilton, Marinship production manager; Carl W. Flesher, US Maritime Commission regional director of ship construction; Ken K. Bechtel, president of Marinship; and William A. "Bill" Waste, Marinship vice president and general manager.

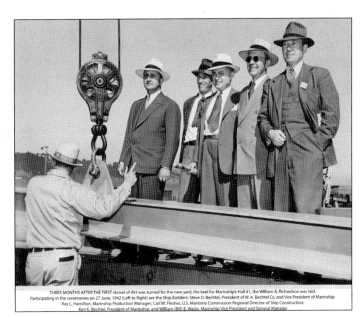

THREE MONTHS AFTER THE FIRST shovel of dirt was turned for the new yard, the keel for Marinship's Hull #1, the William A. Richardson was laid. Participating in the ceremonies on 27 June, 1942 (Left to Right) are the Ship Builders: Steve D. Bechtel, President of W. A. Bechtel Co. and Vice President of Marinship; Ray L. Hamilton, Marinship Production Manager; Carl W. Flesher, U.S. Maritime Commission Regional Director of Ship Construction; Ken K. Bechtel, President of Marinship; and William (Bill) E. Waste, Marinship Vice President and General Manager.

This most historic and incredible panorama shows the condition of the shipyard days after the keel was laid for the *William A. Richardson* and documents how incomplete and primitive the shipyard was when the first keel was laid at Marinship. This panorama was accidentally discovered by the author as he scanned images from a collection of unidentified photographs placed in plastic sleeves in an obscure binder. Only when the photograph was taken out of its sleeve and

This view faces north over Shipway No. 6, several days after the panorama at top was photographed. The *William A. Richardson*'s keel is in place. Shipway No. 5 and the rest of the shipyard are under construction beyond. Pilings beyond Shipway No. 5 reveal where the remaining five shipways will be built. The frame of the Mold Loft and Yard Office is behind the pile driver near the center of picture. The cofferdam is seen starting at middle right and extending to the center of the picture.

laid out to be scanned was it realized both that it was panoramic and what it was a picture of. The extremely rare photograph shows the first few steel sheets of Richardson's keel at far right. Moving counterclockwise, one sees Shipway No. 5 being built next to the sheets of steel being assembled into the first Marinship vessel.

Workmen are scattered all over the shipyard site, hard at work 24 hours a day. A low fog near the top of the image is simulated by the photographer (possibly to prevent the enemy from precisely identifying and locating the shipyard and its components). Note how the tops of cranes and pile drivers fade out. The bottom of Waldo Point is seen at left center, but Mount Tamalpais is not visible. It is right there beyond the shipyard scene in the picture.

The top end of Shipway No. 6 is pictured soon after the keel for the *William A. Richardson* was laid. Close inspection will reveal the very front end of the keel centered on the shipway in front of the mobile crane. Pilings seen to the right of the shipway structure will accommodate the extended shipway needed to accommodate longer and wider tankers that will be built after the last of the Marinship Liberty ships.

Wm A Richardson on Shipway #6 under construction 11 July 1942

The *William A. Richardson* is shown under construction on July 11, 1942.

Subassemblies on Skids during yard construction, probably for Wm A Richardson
Subassembly components made at Calship, sent to Marinship on rail cars, assembled on site as shown here

Subassemblies, probably for the *William A. Richardson*, are pictured here. Subassembly components were fabricated at Calship and transported on railcars to Marinship. These were assembled on the outdoor skids. Hull sections were moved by cranes onto the shipways and welded into place. Work on the congested shipways was minimized, speeding overall ship construction. Many workers could make progress at different locations, and large ship parts were moved to be assembled on the shipway.

This view looks east from Pine Point on August 15, 1942. The *William A. Richardson* is forming on Shipway No. 6 at right. To the left are the *William T. Coleman* on Shipway No. 5, the *William Kent* on Shipway No. 4, and the *John Muir* on Shipway No. 3. Shipway No. 2 is almost ready for the *Philip Kearny*. Shipway No. 1 is being built on the far left. Subassemblies are seen being readied in the foreground.

Looking East from Pine Point on 15 August, 1942

This view looks east from today's Bridgeway over the foundation and initial framing for the Subcontractor Building. The more complete Outfitting Shop is beyond. The Warehouse under construction is at the right. Steel plates are stacked where the Service Building will be constructed to the left of the Outfitting Shops. Shipway No. 6 is just outside the picture on upper left.

View looking East from State Highway (Bridgeway) over incomplete Sub Contractor Building and Outfitting Shop Warehouse under construction is at far right

View looking Northeast from State Highway (Bridgeway) over Compressor Building under construction. Incomplete shipway #6 is above at right. Pile drivers and cranes are building shipways #1 through #5 beyond.

This is a view looking northeast from today's Bridgeway over the Compressor Building under construction. Shipway No. 6 is at far right in background. Pile drivers and cranes are building Shipways No. 1 through No. 5 to the left of Shipway No. 6. Outdoor skids are between the Compressor Building and the shipways. Subassemblies and railcars, which delivered ship components, are seen starting at center left and across the picture.

Looking South from Waldo Point on 15 August, 1942

This view looks south from Waldo Point on August 15, 1942. The Administration Building is at center on left. Behind and to right of the Administration Building is the Plate Storage Yard. The Mold Loft and Yard Office is above the Administration Building. The Plate Shop is taking shape to the right of the Mold Loft and Yard Office. Dredges can be seen removing mud for a ship channel to left of the Mold Loft and Yard Office.

Looking South from Waldo Point on 26 September, 1942

This view looks south from Waldo Point on September 26, 1942. This is the same viewpoint as in the image at top. The Plate Shop is partly operational. Steel, ready to be cut and formed into ship parts, is stored in the Plate Storage Yard. Two small buildings in front of the Mold Loft and Yard Office are the Acetylene-Generation Building No. 1 and the Driox Plant, where oxygen was generated. Today's Bridgeway and Gate 5 roads run from left to right.

Acetylene and Driox Building looking North from NE corner of Mold Loft and Yard Office on 26 September, 1942

This view is looking north from the Mold Loft and Yard Office toward Waldo Point. At center is Acetylene-Generation Building No. 1, one of two such steel and concrete fireproof and explosion-resistant buildings on the shipyard. Acetylene-Generation Building No. 1 is in front of and hiding most of the smaller Driox Plant, where oxygen was generated. The plate storage yard is beyond. A rail line is located to conveniently deliver steel and supplies.

General Shops looking South

This is the General Shops Building, in a view looking southeast from the vicinity of today's Bridgeway at Spring Street. A guardhouse at right services Gate No. 1, and a pedestrian stairway off the picture to the right leads up to the east edge of today's Bridgeway at Easterby Street. The dark cylinder in front of the General Shops is a butane tank that will be placed on a support foundation. Butane was used to heat-form large copper ventilation fittings.

This is the Small Parts Fabrication Area, in a view looking southeast on December 28, 1942. This area is between the Outfitting Shops and the Outfitting Docks. A ramp, seen across the entire picture at center, allowed loaded trucks to drive directly onto the upper level of the Outfitting Docks. Trucks could drive close to ships being outfitted. By design, supplies were efficiently unloaded directly onto ships by mobile cranes, significantly reducing time needed for making deliveries.

In a view looking northwest from Coloma Street, the Training Office Building is pictured under construction. Some classrooms were in this building, along with the training administration offices. The Recycling and Salvage Building is beyond. The Training Office Building was much better built than the salvage building. The Training Building today provides space for artist studios. The site of the Salvage Building, long ago demolished, is now occupied by a school playground.

This is a view looking northeast from the vicinity of Nevada Street, adjacent to Pine Point during peak activity at the shipyard. The Subassembly Building is seen under production. Subassemblies are moved onto the outdoor skids, where they are stored and stacked in order of assembly. Once on skids, they are moved by huge whirley cranes onto the shipways. Shipways and outdoor skids are conveniently located next to each other by design.

A sizeable subassembly piece is seen being fabricated inside the Subassembly Building. Gantry cranes inside the building move subassembly pieces as necessary. Gantry crane tracks are extended beyond the Subassembly Building roof to allow for moving subassemblies outside the building where whirley cranes can pick them up to move them onto the outdoor skids.

This view looks north from a viewpoint adjacent to the guardrail along the east side of today's Bridgeway at the southernmost end of the shipyard. The General Shops, with the Pipe Shop, Copper Shop, and other specialty workshops is at center on the left. Ships tied to Outfitting Docks are seen on the right. Rail lines allow loaded trains to enter the shipyard site to make deliveries. Mount Tamalpais, centered in the far distance, is concealed by fog.

Looking North from South end of Yard on 28 December, 1942

Marinship from Sausalito Hills May 1943

Marinship is seen from Sausalito hills in May 1943. The shipyard is in full production. Most buildings are complete. All shipways are active. Outfitting Docks are at far right. Workers' cars are seen parked along today's Bridgeway at far left. A hillside cut for US Highway 101 is in the foreground at center left. A stiff breeze is blowing welding fumes away across the bay. One can almost hear the din of shipyard activity.

THE PANORAMA FROM WALDO POINT on June 5, 1943 shows many changes, with the Administration Building and Steel Yard in the foreground and the arched-roofed Mold Loft and huge seven bay Plate Shop in the rear.

This is a view of the shipyard from Waldo Point on June 5, 1943. The efficiently stocked Plate Storage Yard provides for full production capability. Many workers' cars are seen parked along today's Bridgeway. Shipyard activities have morphed from construction to ship production. Ships are being built and launched at the average rate of one every 11 days. Marinship is fulfilling its promise to be a most efficient, productive, and modern shipyard.

Two comparison views show shipyard development. This viewpoint is from a photographer's platform suspended from a mobile crane in 1942 and a whirley crane in 1943. The 1942 image shows the *William A. Richardson* on Shipway No. 6. Tankers are being built on the shipways in 1943. At this time, Marinship became known as one of the most efficient and modern, if not *the* most efficient and *the* most modern shipyard operated during World War II.

Three

MARINSHIP IS OPERATED

By early 1943, most of the shipyard had been built. Most effort on the shipyard site was devoted to building ships. There were 20,000 workers, eventually working in three shifts. Ships were started and launched on average at the rate of one every 11 days. The conversion to building tankers instead of Liberty ships had been completed. Innovative shipbuilding techniques had been developed. The complexity and challenges of building tankers compared to Liberty ships had been overcome and conquered. Untrained workers had been recruited to fill seriously deficient labor resources. Workers were trained by master shipbuilders to have limited and extraordinarily proficient skill specialties, similar to the assembly-line technique developed by the automobile industry. The shipyard had been designed to be extraordinarily efficient. Buildings and facilities were grouped together with others where associated tasks and ship products would be convenient to meet future needs. Workers were encouraged to be inventive to develop labor, cost, and time-saving procedures and techniques. Contests for the best invention each week rewarded a chosen worker with a war bond. These inventions were shared with other shipyards. The routine had set in. Everything was going well. The shipyard began to be known as the most efficient builder of tankers. They became the Tanker Champs.

This panoramic view is looking northeast from Pine Point during the peak of shipyard operations. At center, subassembly sections are stacked in the outdoor skids area, ready to be moved onto the shipways by whirley cranes. Shipways are at center starting on right. Three cranes are placed on either side of the outdoor skids—one on the west side (closest to today's Bridgeway) and two between the outdoor skids and shipways. A crane will pick a subassembly section up and move

This is a panoramic view from Pine Point looking southeast during the peak of shipyard operations. This panorama and the panorama at the top have a small common view but a different viewpoint, the part on right at top and the left part of this picture. Smaller subassembly sections are stored at this end of the outdoor skids. Outfitting docks are at center toward the right. Ships launched from a shipway are moved by tugboats to the Outfitting Docks. When first launched, a ship has

it toward the shipways. Heavier subassembly sections require two cranes. A midair transfer of a subassembly section between cranes is possible using special rigging hardware, but the preferred transfer is to place the subassembly section on the ground between the cranes. One or two cranes at the shipways will pick a subassembly section up and move it into place on the shipway.

as many of its components as possible, but there is much more to do before it is finished. At the Outfitting Docks, furnishings and other smaller ship components are installed. This includes pipes, pumps, electrical equipment, wiring, kitchen equipment, bunks, desks, radio gear, and rigging. Small supplies are stored in the Warehouse Building until needed and then moved to the nearby Outfitting Docks.

This is an aerial view looking south. Shipways are the most prominent shipyard component at center. The Outfitting Docks are at top left. The Warehouse is to the right of the Outfitting Docks. The Machine Shop is above the Warehouse, and the Outfitting Shops are below it. Beyond the Machine Shop are the General Shops and the Machinery Storage Building. All of these shipyard components work closely together to outfit a ship. All shipyard components are arranged in proximity to allow for the most efficient shipbuilding possible. Marinship's design was a benefit of being one of the last shipyards built for World War II. Operational experience gained from observing previously built shipyards gave Marinship's designers insight to help arrange the shipyard components to function together as efficiently as possible. It worked. Marinship became known as one of the most efficient and productive shipyards for its size.

A training session is shown. A skilled master shipbuilder, standing, is showing three newly hired managers how to do their jobs. A model helps orient the new hires to ship component locations and to teach ship terminology. The experienced worker trains many inexperienced workers in one of the many shipbuilding skills the master shipbuilder has learned. It was an adaption of the assembly line process developed by Henry Ford.

Newly hired workers are learning how to weld in the classroom. After classroom training, they will gain actual experience in the welding training shop. Welding skills were in very high demand, and few trained welders were available when Marinship began operation. Marinship had to train its own welders from scratch. Men did the welding at first, but the draft took most of them away. Women learned welding and became recognized as the best.

CLASSROOM. Few experienced welders were available when Marinship started building all-welded hulls, so a training school was run in the yard for recruits, many of whom earned while they learned.

This is a view looking northeast showing the Mold Loft and Yard Office. One of the first major buildings completed, it was an important site of operational activities. Offices for shipyard workers were inside along with the Mold Loft. Full-size patterns for cutting the steel plates were built of wood in the Mold Loft. The patterns were moved by workers out the door at the top of the ramp. The patterns were used in the Plate Shop.

This view from Pine Point looks northeast. Many different subassemblies of all sizes and shapes are seen stacked so high that they obscure parts of the shipways. A ship is about to be launched on Shipway No. 5. The congestion in the outdoor skids is because the subassemblies for the next ship to be built are made in advance of need and stored for quick assembly of the next ship.

FOREPEAK FITTING. All of the tanker forepeak sections, extending from keel to uppermost deck, were assembled horizontally to facilitate welding of breast hooks and diaphram plates, then hoisted into position on the ways.

This view shows forepeak fitting. The forepeak is one of the most interesting subassemblies. It is relatively small, but it is a very difficult piece to build properly and accurately. It has complex curves and tight radius bends. It takes lots of time and skill to make one. These were easily made in the Subassembly Building but were challenging to put together on the shipway. In the Subassembly Building, a forepeak could be moved all around as necessary by cranes and specialty jigs used for alignment. Moving into various positions (up, down, left, right, etc.) allowed easy access to tight spaces for welding and forming. It was then moved onto the outdoor skids in advance of need, stored as long as necessary, and lifted into place at exactly the right time. This photograph clearly demonstrates the advantage of subassembly ship construction.

Models were used to help train workers. Marinship had an excellent model shop. Models showed workers how various subassemblies were put together. It is possible that several smaller subassemblies could have been assembled into one large subassembly and then moved in one piece onto a shipway.

A large subassembly is seen being manipulated by a whirley crane. Large subassemblies were created in the outdoor skids area. A subassembly would be moved around as necessary to provide access to surfaces being assembled so that they could be welded in a horizontal position. Welding in horizontal position is much easier and faster than in vertical or overhead positions.

The largest subassembly built at Marinship was the bridge section. The entire bridge was built at the outdoor skids area. Placement of the bridge section was a two-part process. There was the bridge section itself and a bridge support section. These two had to be sequentially lifted by a total of four cranes. Two cranes lifted from adjacent to outdoor skids. Another pair of cranes was alongside a shipway. First, the bridge section had to be lifted and positioned, ready to lift. Then, the bridge support section was lifted into place over the bridge section and onto the ship. A midair transfer between cranes was an extremely dangerous and delicate process. A man was suspended along with a special rigging fixture which allowed the dangerous operation of moving a hook from cranes bringing the load to cranes taking the load.

A group of men and women welder trainees is pictured in the welding training shop. They are being shown the proper use of a welder's "stinger," with an arc-welding rod in contact with pieces of metal held in the hands of an instructor. A stinger is a clamp device connected to a heavy electric cable from a generator. An arc-welding rod is held in the clamp of the stinger.

Tak Kei Ng created this pry bar and angle rest so that it will fit three sizes of stiffeners and eliminate use of blocks to keep angle upright.

Roy Sheldrick demonstrates his burning machine to cut steel disks of any size. With a simple adjustment it cuts the disks with any desired bevel.

Marinship employees were very inventive. A war bond was awarded each week for one of the inventions. Two winners are shown here. Inventions created at Marinship were shared with other shipyards. Marinship became famous for its labor-, cost-, and time-saving inventions. This inventive spirit helped Marinship to become an extremely productive and efficient shipyard.

A tanker is shown being built. Hull plates are being welded near a recently laid keel. A bulkhead is seen beyond. The corrugated bulkhead separated adjacent tanks and prevented sloshing of liquid cargo during transport. Building a tanker was far more complex and time-consuming than building a Liberty ship. Liberty ships had no need for such complex bulkheads. Wood staging allowed access from outside for hull assembly.

The complexity and detail of a tanker is evident. There are miles of pipe in a tanker. There are pumps, motors, wires, and many complex and detailed parts needing fabrication and precision assembly. All of this needed painting. Assembling a tanker required many more man-hours and needed a lot more training and skill compared to what was necessary to build a Liberty ship.

The US Maritime Commission order for Marinship to change production from Liberty ships to tankers was a painful process. Marinship had become very efficient at building Liberty ships, and then it had to change production to tankers. The first tanker was 139 days on the shipways and 66 days at the Outfitting Docks, for a total 205 days. The first Liberty ship took 126 days, and the last one took 60 days.

The USS *Escambia*, Marinship's first tanker, is launched April 25, 1943. Its incomplete status at launch is evident. The bridge is barely in shape. No walls are seen behind the bridge facade. There was a large amount of assembly necessary remaining to be done, which had to be accomplished at the Outfitting Docks. Plumbing, wiring and equipment had not been installed. Marinship had to learn the hard way how to get more of a tanker assembled on a shipway. The shipyard ended up doing this with a passion. Soon, it was launching more tankers per month than any other tanker building shipyard. While incomplete, the *Escambia* is beautifully formed. Marinship tankers became famous for being the best built and most beautiful of them all.

Marinship was among the most decorated of the tanker-building shipyards. Marinship's world-record tanker, the *Huntington Hills*, was built and launched in only 33 days: 28 days on the shipways and 5 days at the Outfitting Docks. This record still stands today. Other tanker-building shipyard records are *Swan Island*, 60 days; *Alabama*, 79 days; *Sun Ship*, 90 days. One of the most famous tankers was the *Tamalpais*. Named for the idyllic Marin County mountain, after a creek was found and named to allow this, it was arguably the most elegant and refined of them all. The *Tamalpais* was the flagship of the Pacific Fleet. The *Tamalpais* was among eight from Marinship tankers assembled for the treaty signing ceremony in Sagami Bay on the occasion of the Japanese surrender. Interestingly, the *Tamalpais* carried water instead of oil.

Four
Marinship Characteristics

During operation of the shipyard, the most important component of it was the workforce. The pieces of steel and the welding gasses and welding rods were simply the media being manipulated by the refined skills of the workers. A finely tuned workforce, directed by competent, experienced, and humane management, was a most important factor. There were inspirations and rewards offered by the few in charge, any of whom would willingly and unhesitatingly step in and personally take over any job whenever the need arose. This included feeding the workers, providing transportation, child day care, housing, and much more. Everyone was motivated and, for the most part, satisfied. Camaraderie was encouraged and developed into a competitive spirit that enhanced and boosted overall shipyard productivity and quality. Not everything was perfect, but it was about the best system that had been created anywhere else on the home front war production effort. While not the first to heed the call to duty to build ships, Marinship was so well designed and so well managed that it became a model and a respected leader. Other shipyards may have been bigger, and other shipyards may have built more ships, but Marinship was the most efficient. Marinship's fame for its low number of man-hours required to build a tanker was envied and became a target for other shipyards to try to match or to outdo. Only a very few of the World War II shipyards have books and movies written about them. Marinship is one of them. This section of the book includes a collection of images that tells these stories using pictures with captions.

"What was it like to work at Marinship? How did you get hired? How much would you get paid? How would you learn the skills you would need, having never worked in a shipyard before? Who would you work with? Where would you live?" All of this would have been on the mind of a prospective employee. This photograph shows Marinship workers on a break or between shifts. The date of the picture is between June 9 and September 30, based on the construction statistics for the ship *Ocklawaha*, a fleet oiler, which is tied up at the dock. Marinship is enduring the change from Liberty ships to tankers. Progress building ships is slow. The *Ocklawaha* took 232 days to build, the second-longest time on record. *Ponaganset* took the longest time to build: 248 days.

Admiral Vickery addresses a lunchtime crowd assembled in front of Shipway No. 4 during his fourth visit to the shipyard. The date is January 27, 1944. The ship on the way behind the platform is *Mission San Luis Rey*, launched January 29, 1944. The admiral had come to Marinship to announce that Marinship might soon be called on to do repair work.

Workers are checking in for their work shift.

Camaraderie was a major feature of working at Marinship. Here, a work crew surrounds an office crew. This image was taken in early 1944. The *Mission San Rafael*, tied at the Outfitting Docks, was launched on March 22, 1944. Many women were working at the shipyard during this time period, but none appear to be ship construction workers in this photograph. Guessing from their clothes, it seems the women are office workers.

This rare image shows prospective workers outside the hiring hall, which was located off the shipyard site for security reasons. Prospective workers were screened for police records before being allowed into the secure shipyard. Some suitcases can be seen. Some job applicants may have just arrived and have not yet found either a job or a place to spend their first night.

Almost anyone could get a job at Marinship. One man hired was partly paralyzed. His left side worked, but his right side did not. He was hired for a job in a toolroom managing distribution and care of precision tools. These workers are all deaf. They are chippers, a job which would deafen anyone with normal hearing. They are spelling "VICTORY" in American Sign Language with their hands.

A machinist is working at a lathe performing a precision task.

MACHINIST. At one time nearly twenty per cent of Marinship's machinists were women, twenty-three per cent of the boilermakers, twenty-six per cent of the drillers and reamers, and twenty-three per cent of the sheet-metal workers.

A female machinist is shown working at a lathe with a turret tool post capable of rapidly and efficiently performing multiple machining functions without stopping work to recalibrate tools. Considerable skill is required to operate this production machine.

An office worker is shown filing plans in a fireproof vault inside the Mold Loft and Yard Office Building. The door to this vault, not visible here, is not unlike the door on a bank vault, with a combination dial and securing levers. Today, one of three such vaults in the building is used as an office for the building manager.

Ten little girls from welding school: Dahla Negro, Vonnie Tipler, Dorothy Tharp, Modena Pike, Lillian Patterson, Lottie Pawley, Frances Angerer, Ella Clarkson, Suzanne Gotts, Ethel Le Beau. Very nice girls!

This photograph is titled *Ten little girls from Welding school*. They are inside the Welding Training Building, learning how to weld. Women welders became known for being better welders than the men. One woman welder said it was easy to spot a man's weld, because it was not as smooth or as accurate as a woman's weld. "Very nice girls!"

Vivian Cochrane, the wife of a Marinship electrical supervisor, was the first woman to hold a job in the construction area of the shipyard. In her fifties when she started working at the shipyard, she was assigned to deliver mail on a bicycle. She was severely resented by a few of the male workers. One offended her by spitting in her face. She went straight to see Ken Bechtel in his office, marching right in without hesitating. Ken asked what was the matter. She told him her story. He listened patiently, asking her to relax while giving her a cup of coffee. He told her to rerun to her office. When she got to her office, her supervisor told her that the man who offended her had been fired and was no longer at the shipyard. Note the mud on her boots.

Joseph James (pictured) was a welder who worked at Marinship. He was also a tenor with the San Francisco Opera and had been a thespian employed during the Depression by WPA-related agencies organized to support minorities. When World War II started, the aid to thespians ended. Joseph needed employment. He learned to weld and was hired at Marinship, but he experienced discrimination from labor union policies. Joseph pressed the issue and was fired from his job. He sued, citing himself and numerous other examples of discrimination. Eventually, the case worked its way up the California court system and was heard by the California Supreme Court, which ruled unanimously in his favor. His personal benefit was minimal, but he did get his job back. The overall effect of his case was far more widespread. The integrated workforce today is one result.

These All-Americans of launching crew keep pace with fast yard tempo.

Men and women are integrated into this launching crew. Several old men are included. The patriotic spirit of workers at Marinship was impressive. Anyone who could would come to do the jobs that had been vacated by the draft.

An all-woman asbestos installation crew is pictured, with an older man as the foreman. Young and old, they all seem to be enjoying themselves.

Reflected in the goggles of this woman burner at Marinship is the bow of the Liberty ship PHILIP KEARNY as it rests on the way preparatory to launching.

This classic and iconic image was obviously posed. There is no clue who the young woman is, but she was undoubtedly selected as a model. Her hair is properly protected with a bandana, as required for women workers. There is a specific procedure for tying the bandana, but there is no rule for what color it is. Today, red and white polka dots are common in Rosie the Riveter reenactment events. The woman pictured is a burner, operating an acetylene cutting torch. The Liberty ship *Philip Kearny*, number 5 of the 15 built and launched at Marinship, readied for launching, is reflected in her goggles. The launch date for the *Philip Kearny* was January 29, 1943. The sponsor for the *Philip Kearny*, the lady who breaks the bottle of champagne, was Nina Elisabeth Palmquist Warren, the wife of the recently sworn in California governor Earl Warren.

A lunchtime crowd enjoys a concert. The band is set up on what looks like a subassembly section and workers are seated on beams used for assembly work in the outdoor skids area. This event is happening in front of Shipway No. 6, with the *William A. Richardson*. The beginning steel assemblies for the *William T. Coleman* are next to Shipway No. 6. This is clearly early during the life of Marinship.

Welding Crew Foreman Kay Daws with her Welding Crew

Kay Dawes, the first female welding crew foreman, is pictured with her fully integrated crew. Dawes was commemorated and honored by William Waste during a special recognition event on her achieving this rare and prestigious status.

S.S. COALINGA HILLS, Launched by Marinship, October 14, 1944

This female crew of burners was nominated for the prestigious honor of cutting the anchor plate that held a ship being launched. An anchor plate holds the ship on the ways until the last segment of steel between predrilled holes is cut. The other half went with the ship. One anchor plate is on each side of a ship. The master of ceremonies calls for each section to be cut sequentially.

General Manager Bill Waste congratulates Kay Daws on becoming the first woman foreman in any skilled craft in any United States shipyard.

General Manager Bill Waste congratulates Kay Daws first woman Foreman in any United States Shipyard

Kay Dawes is recognized for becoming the first female shipyard foreman in all of the United States.

Mrs. Donald E. Reed is obviously having a fabulous time christening the *Mission Santa Clara*. Maybe she swung the bottle a few seconds early. The ship has not yet moved down the shipway. The champagne bottle is decorated with red, white, and blue ribbon covering a brass screen wrap to prevent shattered glass from flying. A cord held by the master of ceremonies links the bottle to the deck of the ship.

The launch crew for the *Mission Santa Clara* is shown on May 18, 1944. The sponsor, her alternate, and the rest of the launch dignitaries pose. Dignitaries might include spouses, close family, the launch master of ceremonies, and the ship's captain. Each sponsor was given a special commemorative booklet with photographs and other statistics. Some of these have been donated by descendants of sponsors and are available for research purposes.

Mrs Edward Winkler, Sponsor for William A Richardson

Mrs. Edward Winkler, wife of a Marinship carpenter, was sponsor for the *William A. Richardson*. This first ship, launched September 26, 1942, was delivered for service on October 31, 1942. She was chosen by lottery. All shipyard workers were issued a numbered ticket. Three of these tickets were selected at random. The three winners selected among themselves by drawing straws to decide who would choose. Edward Winkler won, and he picked his wife.

Madame Wei Tao-wing, representing Madame Chiang Kai-shek, is sponsor for the *Sun Yat Sen*, a Liberty ship launched on March 26, 1943, and delivered on April 17, 1943. China and the United States were allies, and the launch was a highly celebrated event attended by several Chinese dignitaries, California's governor and his wife, and Ken Bechtel.

Madame Wei Tao-wing, representing Madame Chiang Kai-shek, Sponsor for the Sun Yat Sen

Mrs. Joseph Cooper is sponsor for the USS *Escambia*, launched on April 25, 1943, and delivered June 30, 1943. She holds what probably are roses in her bouquet.

Mrs Joseph Cooper, Sponsor for USS Escambia

Mrs J. H. Pomeroy, Sponsor for the Mission San Carlos

Mrs. J.H. Pomeroy is sponsor for the *Mission San Carlos*, launched February 12, 1944, and delivered April 15, 1944. She holds the traditional roses in her bouquet.

Mrs. Ralph K. Davies is sponsor for the *Mission Santa Ynez*, launched December 19, 1944, and delivered March 13, 1944. She holds the traditional roses in her bouquet.

Mrs. Ralph K. Davies, Sponsor for the Mission Santa Ynez

Kathleen F. King is sponsor for the SS *Andrew White* on January 28, 1943, and delivered on February 27, 1943. Her bouquet has acacia. Though not confirmed, she is probably holding roses outside of the photograph's frame.

Sally Rand is shown with some lucky shipyard workers. Celebrities toured the shipyards every week or so, usually during lunch, to boost morale of workers and to sell war bonds. Workers who were not able to attend could listen over the Marinship in-yard broadcast loudspeaker system.

SALLY RAND AND SHIPBUILDERS. Many celebrated entertainers visited Marinship to pep up the workers and stimulate war bond sales.

Susan Hayward entertains shipyard workers.

Bing Crosby entertains a lunchtime crowd on a windy day. Note that the cloth wrapped around the microphone as a wind muffler. At least one starry-eyed lady is among the group at front center. The photograph is titled *Der Binger Croons*.

One can only guess what the event was this noon hour. Was it visiting dignitaries, a boxing or wrestling match, or a band concert? Situated between the Outfitting Docks and the Warehouse, shipyard workers are perched on everything from rooftops to stacks of wood pallets. Motivational events, contests, and reward ceremonies were important for keeping up morale among the workers.

This event appears to be a well-attended US Army Band concert in front of Shipway No. 3.

A US Navy Band concert is shown between the Outfitting Docks and the Warehouse. In the lower-left corner of the crowd near camera are either civilian attendees or office workers.

Marching bands and parades were a very popular lunchtime event. The marching band and escorts are Navy.

An Army band from the Presidio is the attraction in this parade. The event appears to be celebration of a Tanker Champ award.

Royal Saudi visitors came to Marinship. His Royal Highness Prince Faisal bin Abdulaziz is the honored guest on this occasion of a trial run. The prince is seen in Ken Bechtel's office being presented with a "Pictorial Highlights of Marinship" commemorative portfolio of photographs. In addition to the portfolio, Prince Faisal was given a model of a tanker. The prince was in San Francisco for the opening ceremonies of the United Nations. He was interested in seeing tankers, which were transporting oil from his country for the war effort. He brought his entourage to Marinship and was given a royal tour.

The Saudi royal entourage is escorted up the gangway onto the *Tamalpais*.

The Saudi royal visitors have free run of the ship. Security guards are strategically located among the crowd.

The prince inspects shipyard facilities. The workers are fascinated and thrilled by the royal entourage.

Saudi royalty poses with Steve (left) and Ken Bechtel inside a cabin on the ship.

The event was catered inside a facsimile of an Arabian tent set up on the deck of the ship.

Musicians entertain the guests as they have their lunch.

The launch of the *Mission San Francisco* is pictured here. Probably the most celebrated launching of the Marinship shipyard, this was also the last. The invocation was by Dr. Lynn Townsend White, a professor at the San Francisco Theological Seminary in San Anselmo. Ken Bechtel recounted nostalgically how Marinship had won top honors in tanker construction. Irish soprano Agatha Turley sang "Beyond Blue Horizon." Special congratulations were conferred by Carl Flesher, former regional director of the US Maritime Commission. Bill Waste, vice president and general manager of Marinship Corporation, introduced the sponsor and guests. Armand Girard, baritone, sang "Glory Road" and other favorite numbers. The crowd insisted on "Short'nin' Bread." A blimp floats behind the crane at right. The scene shown in this iconic photograph, titled *Marinship's Arch of Triumph*, will be celebrated by a shower of confetti as soon as the ship starts to move down the shipway.

7. The *SS Mission San Francisco* was the 93rd and last ship launched from Marinship's ways. The ceremony was held on September 18, 1945. In 1957, this ship met her demise on the Delaware River. At the time, her tanks were empty of oil, but contained a residue of gas fumes. When her hull was struck by a small freighter, a violent explosion destroyed the vessel.

The SS *Mission San Francisco*, a Mission-class tanker, was the 93rd ship launched from Marinship's ways. The launching ceremony was held on September 18, 1945. It was delivered, ready for service, on October 11, 1945. The keel was laid May 5, 1945, one month before D-day (June 6, 1944). It was on the shipways for 136 days and at the Outfitting Dock for 23 days—a total of 159 days. It was not the fastest-built tanker (*Huntington Hills* was at 33 days) or the slowest (*Ponaganset* took 248 days). Ship construction was delayed by dwindling availability of supplies and labor. In 1957, the ship met its demise on the Delaware River. At the time, the ship's tanks were empty of oil but held a residue of gas fumes. When the hull was struck by a small freighter, a violent explosion destroyed the vessel.

24 The ceremonial launch of the SS Kern Hills (Hull 77) was a big event because it marked the 1 million tons of tankers built at the Marinship yard. For the launching celebration, the ships had elaborate bow paintings such as this one of the bear seen here. When the ships went to the Outfitting Docks these graphics were painted over with various tones of grey

The ceremonial launch of the *Kern Hills* (Hull 77) on March 27, 1945, was a big event because it commemorated one million tons of tankers built at the Marinship shipyard. For the launching celebration, the ship had elaborate bow art showing the mascot Marinship bear.

87

The Mission-class tanker SS *Santa Clara* is launched on May 18, 1944.

SEVERAL THOUSAND EMPLOYEES and their families crowd about Way 6 to watch the SS Escambia, Marinship's first tanker, take to the water April 25, 1943.

Marinship employees and their families watch the launch of the SS *Escambia* on April 25, 1943. The SS *Escambia* was the first tanker to be built and launched at Marinship.

A large crowd is gathered to watch the launch of the SS *San Francisco* on September 18, 1945, photographed from inside the support structure of a whirley crane.

The ways are greased to help a ship slide down into the bay during its launch.

Measurements are made to ensure precise conformance to plans and specifications.

A launch crew removes blocks placed to support a ship during construction. A ram is carefully and methodically operated by a skilled crew prior to launching.

An anchor plate is cut on commands made by the master of ceremonies during a ship launch. Two skilled burners will flame-cut the steel sections marked with numbers between predrilled holes as the master of ceremonies directs them. At the command "Burn one," two burners will cut through the steel of one section on each of two anchor plates, one on each side of a ship, etc.

Army Barges were built at Marinship during 1945 on Ways to the NorthEast of the main yard. Barges were to transition from cargo carriers to invasion craft. First built using rivets but assembly time took too long and was changed to welded construction. A total of 19 barges were built. They were intended to be used to invade Japan. The surrender of the Japanese ended barge construction and the barges were never used as intended.

Army barges were built at Marinship during 1945 on and near Shipway No. 1 after the staging used to build ships had been cleared away. Barges were first built using rivets, per Army specifications, but assembly took too long. Speedier welded assembly was substituted for rivets. A total of 19 barges were built. Barges were to transport cargo and then transition into invasion craft. Japanese surrender ended barge construction. Barges were never used as intended.

The barge *Daisy* is pictured being launched. Daisy Hollingsworth, Ken Bechtel's personal secretary, was the sponsor. Several Marinship launching traditions were not observed during a barge launch. Instead of arriving in an automobile, the sponsor arrived on a bicycle built for two with the master of ceremonies on the rear seat. The corsage was made of daisies instead of roses. The gift watch was tin, not gold. Water was in the bottle instead of champagne.

Bow art was a tradition only at the Marinship shipyard. Commemorative subjects were painted on each hull for decoration. After a ship had been launched and it was being finished at the Outfitting Docks, the art was painted over with shades of grey. The US Maritime Commission would not allow the art to remain, because it could be used by enemy forces to identify the ship.

This bow art depicts support for Gen. Dwight David "Ike" Eisenhower, the Supreme Allied Commander in Europe. Ike would go on to become the 34th president of the United States.

This bow art was for the SS *Mission Buenaventura*, launched May 28, 1944.

Bow art subjects were intended to remind workers that their individual actions could have unintended consequences contradictory to the home front national defense effort.

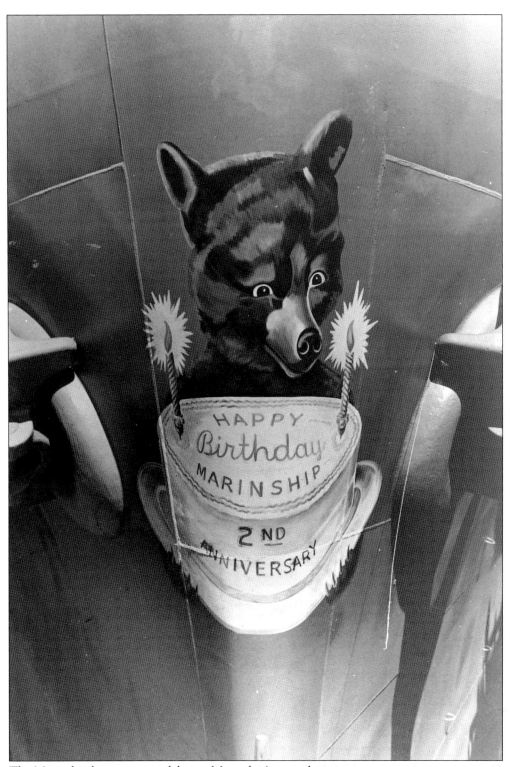
The Marinship bear mascot celebrates Marinship's second anniversary.

The Marinship bear mascot commemorates Texas, painted on the SS *Concho*, launched July 25, 1945.

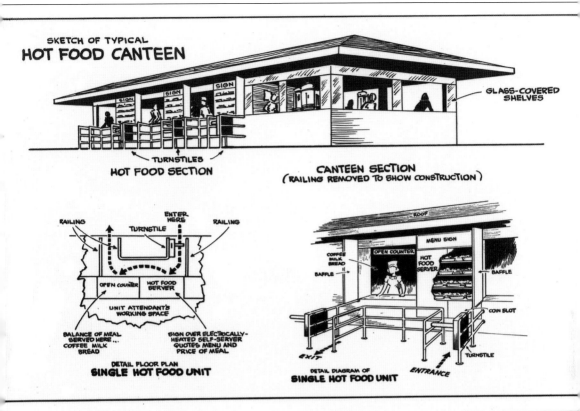

Food for Marinship workers was an important management challenge. At first, there was little to offer. Local merchants, today's food trucks, filled the need. From this, a more organized effort evolved. Caterers made sandwiches off-site and delivered them to established locations on the shipyard. As it developed from a construction site into a shipyard, a more permanent solution was needed. A Canteen Building was built across today's Bridgeway from the Administration Building. It could seat 300 people, but it was too remote for shipyard workers. Canteen workers made box lunches, which were delivered to established distribution facilities located at population centers on the shipyard. A turnstile arrangement was implemented. If a worker wanted turkey, he would put his coins in the turnstile labeled "Turkey." He proceeded to the window and had his box lunch a few seconds later.

MARINSHIP CAFETERIA

The main Canteen Building is shown. Today, it is used as a private school. It could seat 300 people. Only administration-building employees could conveniently eat at the canteen. It was too remote for shipyard workers on their lunch break. Canteen workers made hot box lunches to be delivered to distribution facilities at population centers on the shipyard. They were very popular.

FAST-FEEDING UNIT. Six canteens, each named for a ship, were placed at population centers throughout the yard to provide workers with complete hot meals, in three seconds, at fifty cents apiece, or less, during each shift. Two other units, previously erected, also dispensed food.

The largest of the shipyard's on-site facility Satellite Canteen Buildings, William A. Richardson, is seen in action at lunchtime. There were five other similar but smaller Satellite Canteen Buildings on the shipyard. Each was named after a ship built at Marinship. Each hot box lunch included two sandwiches, pastry, salad, and fruit for 35¢ and coffee or hot soup for 5¢ more, served at the rate of one every three seconds or less.

Housing was a huge problem. There were many workers and not enough housing. The housing shortage was relieved by building Marin City. Planned in three days and built as public housing by the US government in weeks, it quickly became Marin's second-largest city. Monthly rent ranged from $29 for a one bedroom apartment to $43.50 for a six-room unfurnished house. Utilities and hospital care were included in the rent.

54 Marin City was built by the U.S. Government to house Marinship workers, many recruited from the Deep South. It was planned in three days and built in weeks, becoming by the end of 1943 the second largest town in Marin County. Rents ranged from $29.00 a month for a one bedroom apartment to $43.50 per month for a 6 room unfurnished house. Utilities and hospital care were included in the rent.

Two locations for housing are shown. The area today known as Marin City, west of US 101, is at the top. This was the family housing area with school, market, and community center. Below are the single men's and women's barracks. Meals at the barracks were served at cafeterias. Today, the barracks site is occupied by a school to the north of Nevada Street. The buildings are mostly gone now.

The newly completed apartments were eagerly inspected during a well-publicized open house. All available housing was occupied as soon as possible.

Several thousand persons inspect the 700 apartments at Marin City

The apartments were clean and comfortable. While welcome and very enjoyable during the war, they were originally cheaply and quickly built. They did not meet the standards for permanent housing expected by returning GIs. It was not too long before they were demolished as substandard housing.

Commuting to work during war-rationing America times was a huge problem. Each person was allocated something like four gallons of gasoline per week. Tires, and even shoe leather, were also severely rationed. Carpooling was the first remedy.

An organized commuter transportation system was soon implemented. The federal government acquired buses and contracted with operators to plan routes and provide affordable scheduled transportation for shipyard workers. Pictured is a shift change near the bottom of Easterby and Spring Streets at today's Bridgeway. Before buses were available, a trailer transportation system was provided.

Shiny new buses were made available for commuters. All workers used them. Shipyard workers, office workers, and managers welcomed the transportation opportunities made available to them. A few of these buses are today maintained by historical societies. Some may be chartered for reenactments and other historical events. Hollywood movie producers are frequent clients.

A Southern Pacific Railroad ferry, mothballed because the Golden Gate and Bay Bridges eliminated the need for them, was refitted and put into service. There were two ferry terminals in San Francisco, one to the south of the Ferry Building and one at Hyde Street Pier. The ride cost 15¢. Workers could sign up for welding classes on the ferries and learn how to weld while they commuted.

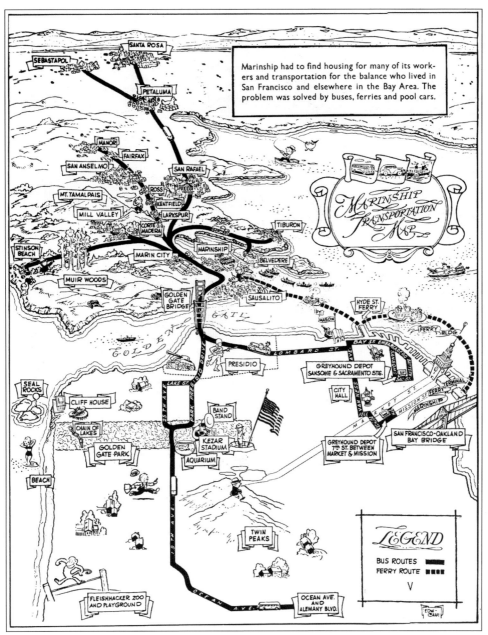

This poster, typical of the times, advertised the commuter system to shipyard workers to help them to get to work. The bus routes extended from the far south at Ocean Avenue and Alemany Boulevard, to Santa Rosa and Sebastopol on the north and Stinson Beach on the west. Downtown San Francisco was served by routes to two Greyhound depots, one at Sansome and Sacramento and the other on Seventh Street between Market and Mission Streets. Ferry routes are also shown. As many remote and metropolitan locations as possible were accommodated to encourage shipyard workers to come and work at Marinship. There was a special edition of the *Marin-er* published on August 5, 1944, to announce available jobs at Marinship. This graphic was part of that edition of the *Marin-er*. As many as 3,000 workers used the bus and ferry transportation system each shift.

FERRY SERVICE TO MARINSHIP

The ferry terminal is located at what today is the Clipper Yacht Harbor and Marina at the end of Harbor Drive. The pilings for the ferry landing endured well into the 1960s. A boat launch ramp occupies the site today.

The on-site bus terminal offered efficient transportation for shipyard workers. The many destinations available are identified for individual buses. Located near the shipyard ferry terminal, the buses are lined up discharging their passengers at shift change time. Buses were in great demand during the war. Many different styles of bus can be seen. All were well made, even if they were older models. Some are classic and rare antiques today.

Family Day was a regular event at Marinship. Launchings were the usual attraction. One wonders who is in this portrait. Is this a family? Are they brothers, sisters, spouses, cousins, uncles and aunts, or grandparents?

Off the shipyard site, workers had many opportunities available to them in one of the most desirable living locations anywhere. Many decided to stay after the war ended.

The pleasures of living in Marin county, one of the finest residential areas in the nation, were one attraction for Marinship workers. Here

A birthday party at the Child Care Center is pictured.

This children's event on the shipyard includes entertainment by a clown.

The Medical Clinic on the shipyard site treated workers promptly. Eye care was one important function often provided here. Cuts, bruises, broken bones, colds, and fevers—all were taken care of conveniently and at no cost. The rationale was to keep employees healthy and get them back to work as soon as possible.

CLINIC. At the Marinship Hospital patients were treated as they would be treated by their family physicians, but without frills. Treatment was given first, questions asked afterward. There were as many as 500 cases a day during the peak, but serious injuries were comparatively few.

A tanker assembly crew pauses for a group portrait. Most are white men. A handful of them are minority workers, and a dozen or so are women, mostly welders. Crews such as these were generally assigned to a shipway. They worked together daily to get things done well and efficiently. Here, they seem to be posing in trade groups. The welders (wearing helmets) are on the bottom level.

The Kangaroos were a 10-day crew. The men always worked together and probably had a specialty, such as assembling bulkhead sections. They were assigned to do a job and expected to finish it in ten days or less. There were all sorts of 10-day crews. They could be sent to a problem area to get it finished quickly. During the peak of operations, there were 50 or so 10-day crews.

This is a rigging crew. Rigging was a dangerous job. Each team member had to trust his coworkers, and each coworker had to support the others. A slight mistake, and some huge, heavy subassembly could slip or fall and cause the immediate end or a serious permanent injury for one of them.

Teamwork was a major factor contributing to Marinship's success. The day shift would compete with the night shift or the swing shift. One crew would accomplish a task, and the other crews would try to outdo it. This helped Marinship to be one of the most productive of all the shipyards of its size.

This is probably a 10-day crew at work on a forepeak subassembly in the Subassembly Building.

FOREPEAK ASSEMBLY. In the Sub-Assembly Shop men and women swarmed about like bees in a hive, building up section after section—fitting, flanging, burning, welding.

Here, an anchor windlass is being lifted to be placed on deck. Historically important, this windlass was rushed by truck from Portland, Oregon, to be placed as the last thing necessary to achieve completion before launch of the world-record-holding, fastest-built Marinship tanker, the *Huntington Hills*.

ANCHOR WINDLASS. The last piece of machinery to be hoisted aboard the record-breaking tanker *Huntington Hills* was an anchor windlass brought by truck on a non-stop run from Portland, Oregon, and delivered the afternoon of the launching.

In this portrait of an office party, Richard Grambow, the youngest member of the Marinship management team and Marinship's naval architect, is at far left.

Members of the Marinship basketball team are pictured here. It looks like there are two US Navy Shore Patrol guards sitting in the top row of the bleachers. There was no basketball court on the shipyard. The location where this picture was taken is not known.

Gray Davis, a shipyard worker and an accomplished artist, is pictured. His paintings immortalize a few of the aspects of Marinship life. His artistic license has placed a pickup truck in the foreground of this painting.

This posed image documents an official artist at work immortalizing a ship onto canvas. Interestingly, he has seen fit to place a car in the foreground of his painting!

Gray Davis is shown with the three pieces of his Marinship art that are part of the Bay Model Marinship exhibit. It is believed that the artist personally donated these paintings for display. At left is a night scene on a shipway. At center is a Machine Shop scene showing a large solid cast steel stern section. On the right is a female welder.

Marinship had an excellent model shop. A model of a tanker is shown. The large scale and detail of these models allowed study of proposed changes to be made and trained workers in more efficient assembly strategies. The fate of these models in not known.

MODEL TANKER. The construction of tankers was explained to student welders by means of sectional scale models. The models were sometimes used on the ways, too, as aids to the working out of rigging problems and improved methods of erection.

Workers in the Electrical Shop are precision-bending a large piece of electrical conduit.

A worker in the Plate Shop uses an automated machine burning rig to make three precision cuts at one time.

An accountant or Warehouse administrator is keeping track of production or inventory.

The blueprint machine is shown. Thousands of feet of blueprints were needed daily. Every ship had to have its own complete set of blueprints to accompany it as it was operated. And it took many separate sets of blueprints—at least one partial or complete for each trade specialty—to build a ship. Note the large vent hood, which helped to remove the ammonia smell of the developer used to make blueprints.

Two ladies are pictured learning about the use of acetylene-flame cutting torches. While not a difficult or strenuous task, using a "torch" requires finesse and delicate dexterity. Women were ideally suited for the task. The lady on left is properly dressed with a leather jacket; the lady on right is wearing what looks like denim. The leather will endure, but the denim will soon be full of holes from hot slag sparks.

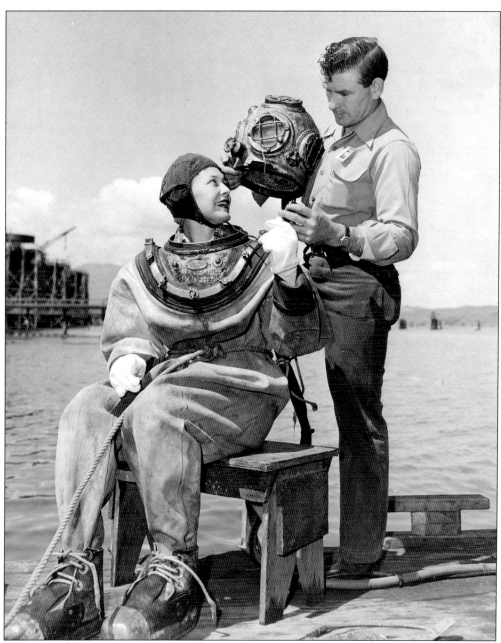

This intriguing picture shows an unidentified female diver being assisted with suiting up. She seems to be enjoying the assistance being offered to her. Not much is written about what divers did at Marinship. Speculating from what divers do today, it was probably about the same. Maintenance of docks would be common. And perhaps fixing a leak in a hull after launch. Whatever was done, it was dangerous and highly specialized. Underwater construction was a job women were not expected to want to do; nor were they ideally suited for it. However, with no men around to do it, someone had to step up to the plate and take the challenge.

Broadway and Hollywood stars inspired emerging Marinship talent to entertain their coworkers. The top of this *Marin-er* article features world-class entertainers. From left to right are Marian Anderson, whose beautiful contralto voice inspired thousands who heard her; Bill "Bojangles" Robinson, world-famous tap dancer; (accompanying Robinson) the Three Peppers, masters of the old swingeroo, including pianist Oliver "Toy" Wilson, guitarist Bob Bell, and bassist Walter Williams; soloist Todd Duncan, gifted baritone who portrayed Porgy in Gershwin's *Porgy and Bess*; and soprano Gertrude Farmer. Their Marinship compatriots would take any opportunity to entertain at Marinship. No recordings are known to exist of any such performances. Actually, any record of a performance by (or a feature story about) minorities is rare. There are only a few in all editions of the *Marin-er*.

At Marinship, there was the *Marin-er*, a biweekly publication announcing news and sharing human interest stories for the shipyard workers. Lilian Banks is featured in one part of a rare feature article about minority workers and their contributions to the war effort.

This is a section of a page from the *Marin-er* featuring the stories of several minority workers who heard the call to come and be a part of the Marinship labor force. They came from all parts of the country and all walks of life to be shipyard workers. They were enthusiastic about their opportunities in the shipyard.

VICE ADMIRAL HOWARD L. VICKERY

Admiral Vickery was in charge of it all. Although Admiral Land sent the telegram pleading for a shipyard to be built, it was Admiral Vickery who reviewed the proposal to build and who oversaw construction and operation of Marinship. The admiral visited Marinship on several occasions to inspire, cajole, and congratulate. Not everything was going well all the time. A visit from the admiral was often one that was feared. He would make surprise visits to see what the problem was and to do what he could to remedy whatever it was that concerned him. He had an extremely stressful job. The stress got to him. A heart attack in September 1944 laid him up in bed. He retired before the end of the war, living in Palm Springs until he died there at age 53 in 1946.

Admiral Vickery is pictured while touring Marinship during construction. On his right is Ken Bechtel. On his left is Robert Digges, administrative manager at the shipyard. They are surrounded by a cadre of security agents. Security at the shipyard was always tight, but on the occasion of an Admiral Vickery visit it was stepped up a notch.

Here, the admiral is on another visit during shipyard construction. On his right is William Waste, vice president and general manager of the shipyard. On his left is Ken Bechtel. He was most likely visiting to present the "M" for "Merit" flag to Marinship for "contributing" more during 1942 in ship production than any other six-way shipyard started at the same time.

This iconic nighttime view of the shipyard shows it during peak operations. Not everyone was running around busily making ships. This photographer took time to reflect on the situation and captured this view for all today to appreciate the historic scene. Without this idyllic view to show what it was like, readers today might not understand the determination and effort Marinship's workers contributed toward achieving world peace.

Marinship is shown in full operation at night. The shipyard operated 24 hours a day and brilliantly lit up the night sky. It is calculated that at its peak of operation in February 1945, the shipyard required 11,200 kilowatts of power. It was so much power, in fact, that a new 55-mile-long, high-voltage electrical power transmission line was needed from Sausalito to the Vacaville substation on the California power grid.

It happened here in Sausalito. There was nothing like it when it was in operation. There will be nothing like it ever again. When it arrived, it was not certain the City of Sausalito wanted it. When it was here, Sausalito was glad. It changed Sausalito forever. This view is looking north toward Mount Tamalpais from inside the Plate Shop.

When the shipyard stopped building ships at the end of the war, the Maritime Commission suggested that the W.A. Bechtel Company operate the yard. In consideration of the other shipyards also in the same status, the W.A. Bechtel Company, probably wisely, declined the offer. The W.A. Bechtel Company, instead, suggested that the US Army Corps of Engineers take the shipyard over and use it as a base of operations for its Pacific Island Reconstruction Program. On May 16, 1946, the Army Corps of Engineers took over the plant facilities. The corps kept some for its own use and sold the rest off to the highest bidder. Some years later, William Waste, general manager of the Marinship shipyard, wrote nostalgically, "There is nothing left of Marinship but some rusty steel, empty Ways and Docks, and dark, damp buildings."

The Marinship flag was inspired by the California state flag, which features a California grizzly bear. Marinship was a proud California corporation, and it proudly associated itself with the state of its origin. Thus, a bear became the unofficial Marinship mascot—or perhaps it actually was the official mascot.

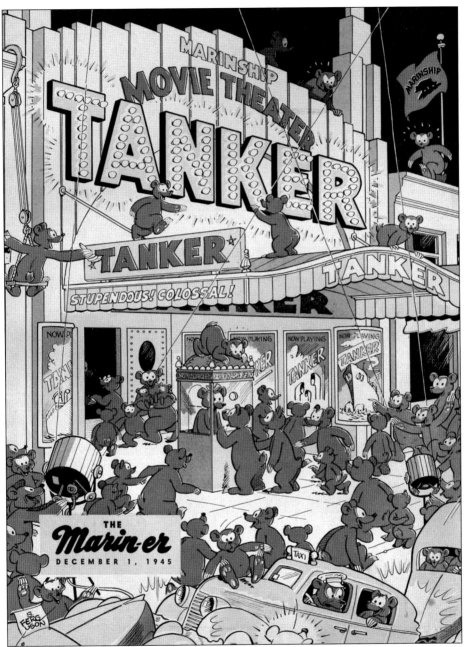

Tanker is a movie that was produced by the Marinship Corporation near the end of the war. It was a gift to the workers of Marinship for a job well done. It boosted their morale at a time when they were recovering from losing their high-paying jobs. The movie was shown in first-rate theaters and was attended by workers and celebrities. One may wonder why there are bears in the poster. The bear is the Marinship mascot. The movie documents the Marinship event in history. It tells the Marinship story from a perspective in time that all of the workers experienced. They lived it. They worked in it. They were part of it. They enjoyed it. They were proud of it. The *Tanker* movie can be viewed online (https://archive.org/details/chi_00005) and has a running time of 47 minutes.

Discover Thousands of Local History Books
Featuring Millions of Vintage Images

Arcadia Publishing, the leading local history publisher in the United States, is committed to making history accessible and meaningful through publishing books that celebrate and preserve the heritage of America's people and places.

Find more books like this at
www.arcadiapublishing.com

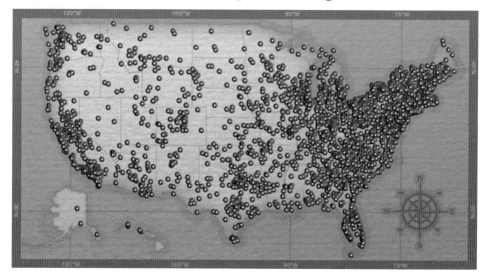

Search for your hometown history, your old stomping grounds, and even your favorite sports team.

Consistent with our mission to preserve history on a local level, this book was printed in South Carolina on American-made paper and manufactured entirely in the United States. Products carrying the accredited Forest Stewardship Council (FSC) label are printed on 100 percent FSC-certified paper.